NOFA

Organic Principles and Practices Handbook Series
A Project of the Northeast Organic Farming Association

Organic Soil-Fertility and Weed Management

Revised and Updated

STEVE GILMAN

Forewords by Fred Magdoff and Lynn Byczynski
Illustrated by Jocelyn Langer

CHELSEA GREEN PUBLISHING
WHITE RIVER JUNCTION, VERMONT

Originally edited by Jonathan von Ranson with
assistance provided by volunteer NOFA farmers
Julie Rawson, Lynda Simkins, Donna Berlo,
Rich Williams, Beth Henson, and Chris Holopainen.

Organic Soil-Fertility Management and *Organic
Weed Management* were first published in 2000 by
NOFA/Mass., 411 Sheldon Rd., Barre, MA 01005,

First Chelsea Green editions published 2002.

Chelsea Green revised and updated combined edition
published 2011.

Editorial Coordinator: Makenna Goodman
Project Manager: Bill Bokermann
Copy Editor: Cannon Labrie
Proofreader: Helen Walden
Indexer: Peggy Holloway
Designer: Peter Holm, Sterling Hill Productions

10 9 8 7 6 5 4 3 2   16 17 18 19

Library of Congress Cataloging-in-Publication Data
Gilman, Steve.
    Organic soil-fertility and weed management / Steve Gilman ; forewords by Fred Magdoff and Lynn Byczynski ;
illustrated by Jocelyn Langer. -- Updated and rev.
        p. cm. -- (Organic principles and practices handbook series)
    "A Project of the Northeast Organic Farming Association."
    Originally published: Barre, MA : NOFA/Mass., 2000.
    First Chelsea Green ed. published 2002.
    ISBN 978-1-60358-359-6
    1. Soil management. 2. Soil fertility. 3. Organic farming. I. Langer, Jocelyn. II. Northeast Organic Farming
Association. III. Title. IV. Series: Organic principles and practices handbook series.

S591.G56 2011
631.4--dc22

                                    2011002169

Chelsea Green Publishing
85 North Main Street, Suite 120
White River Junction, VT 05001
(800) 639-4099
www.chelseagreen.com

Best Practices for Farmers and Gardeners

The NOFA handbook series is designed to give a comprehensive view of key farming practices from the organic perspective. The content is geared to serious farmers, gardeners, and homesteaders and those looking to make the transition to organic practices.

Many readers may have arrived at their own best methods to suit their situations of place and pocketbook. These handbooks may help practitioners review and reconsider their concepts and practices in light of holistic biological realities, classic works, and recent research.

Organic agriculture has deep roots and a complex paradigm that stands in bold contrast to the industrialized conventional agriculture that is dominant today. It's critical that organic farming get a fair hearing in the public arena—and that farmers have access not only to the real dirt on organic methods and practices but also to the concepts behind them.

About This Series

The Northeast Organic Farming Association (NOFA) is one of the oldest organic agriculture organizations in the country, dedicated to organic food production and a safer, healthier environment. NOFA has independent chapters in Connecticut, Massachusetts, New Hampshire, New Jersey, New York, Rhode Island, and Vermont.

This handbook series began with a gift to NOFA/Mass and continues under the NOFA Interstate Council with support from NOFA/Mass and a generous grant from Sustainable Agriculture Research and Education (SARE). The project has utilized the expertise of NOFA members and other organic farmers and educators in the Northeast as writers and reviewers. Help also came from the Pennsylvania Association for Sustainable Agriculture and from the Maine Organic Farmers and Gardeners Association.

Jocelyn Langer illustrated the series, and Jonathan von Ranson edited it and coordinated the project. The Manuals Project Committee included Bill Duesing, Steve Gilman, Elizabeth Henderson, Julie Rawson, and Jonathan von Ranson. The committee thanks SARE and the wonderful farmers and educators whose willing commitment it represents.

ACKNOWLEDGMENTS

Thanks to Carl Reidel, Amy McMillon, Remi Gratton, Carol Dunsmore, Gwyneth Flack, Steve St. Onge, and Doug Flack for the editorial help. Special thanks also to Nat Bacon and Lisa McCrory, and to the many farmers and veterinarians who provided me with information and inspiration.

CONTENTS

PART 2: Organic Weed Management

PART 1

Organic Soil–Fertility Management

Foreword to Part 1

The primary management goal for organic farmers should be to build healthy soils on their farms. A healthy soil is the foundation for healthy plants and, therefore, a healthy farm. Creating a soil that is healthy—biologically, chemically, and physically—is not a simple task and cannot be accomplished by using only one or two practices. The creative management of the farmer, and not a cookbook recipe, is the path to healthy soils. There are, however, certain guidelines, approaches, and rules of thumb that can provide a framework for farmers.

Soil organic-matter management is at the heart of soil fertility for organic farmers. Organic matter influences almost all soil properties. It provides food for soil organisms, and, during the decomposition process, nutrients are released from organic matter, and sticky substances that help form soil aggregates are produced. Better plant nutrition and physical condition of the soil—with better water infiltration and storage—promote healthier plants. Many hormonelike substances are produced in soils that stimulate plant growth.

A healthy soil must have plentiful supplies of organic matter added regularly: crop residues, cover crops, manures, composts, and so forth. If practices that promote healthy soils are routinely carried out, most of the problems relating to nutrient supplies for plants will be taken care of along the way. Higher cation-exchange capacity, caused by increasing organic matter, allows the soil to hold on to potassium, calcium, magnesium, and ammonium more effectively. Lots of biologically active organic matter—in addition to promoting a diversity of soil food-web organisms—means that nitrogen, phosphorus, sulfur, and other elements become available to crops as soil organisms feed on residues of crops, cover crops, manures, and composts.

As pointed out in this handbook, there are no magic bullets to creating healthy and fertile soils. Testing soils regularly helps to identify, and then remedy, potential problems of soil acidity or low levels of nutrients.

Attention to basics, and the knowledge gained by careful, firsthand observation—the farmer's "footprints" on the soil—are the answer.

FRED MAGDOFF
May 2002

Fred Magdoff is professor of plant and soil science at the University of Vermont and regional coordinator of the USDA's Northeast SARE (Sustainable Agriculture Research and Education) Program. He is the coauthor of *Building Soils for Better Crops* (Sustainable Agriculture Network, 2000), and coeditor of *Soil Organic Matter: Analysis and Interpretation* (American Society of Agronomy, 1996) and *Hungry for Profit: The Agribusiness Threat to Farmers, Food, and the Environment* (Monthly Review Press, 2000).

Introduction to Part 1

Everyone knows what soil is. Beneath our feet it's the geological residue known as earth, ground, terrain, land, acreage, real estate, and good old terra firma. Used in producing our food, it's called farmland, field, pasture, patch, bed, and garden. Under our fingernails or tracked across the kitchen floor it's nothing but dirt, soil in the wrong place.

Beyond its nitty-gritty geological components, soil is a concept as much as anything else. Its varied perception has evolved over millennia through a mixture of farmers' practical experience and the latest scientific investigation. Each gain in understanding of how soils function helped define the agriculture of the time. Although farmers had long used manures, ground bones, ash, and plant wastes to increase crop production, the essential "mineral" nutrient basis of plant growth wasn't grasped until Justis von Liebig's research in Germany in the 1840s, which substantially altered the viewpoint of the nature of soil and plant nutrition.

Such was Liebig's stature as the preeminent chemist of his time that his discoveries dominated the thinking of soil scientists well into the twentieth century. The subsequent industrial embrace of his focus on the chemical properties of the soil still holds sway today. He determined that plants utilize nutrients in an inorganic, mineral form and supposed this occurred through chemical reactions in the soil. Liebig's "law of the minimum" demonstrated that the diminishment or lack of any one essential nutrient, even if only a trace amount was needed, could limit the growth of plants.

While Liebig was not necessarily to blame for the extent of the Big Lie that followed—that plant growth is solely a function of available soluble chemical nutrients—his work provided the basis for the huge commercial chemical fertilizer industry that followed. Ground bones were long used by farmers as a fertilizer source, for example, but Liebig treated them with sulfuric acid to increase the solubility of their phosphorus content. Today's superphosphate is made the same way.

First called "artificial manures," the early fertilizer formulations combined nitrogen compounds (commonly urea, ammonium sulfate, and calcium cyanamide) with available natural waste materials such as straw, corn stover, or spoiled hay. While they were designed to supplement the barnyard-variety manures, farmers of the day were initially not convinced. They felt that such chemical manures would only stimulate the crop and eventually exhaust the soil.

The other part of the equation—the integral role soil microorganisms play in mineralizing such nutrients from organic matter as part of the decay process—remained unknown until a quarter century later. This discovery by Winogradsky was based on the work by Pasteur that opened up a new microscopic world of soil inhabited by bacteria and other organisms. In 1882, Darwin provided a view into its complex biological nature in his study of earthworms. A new school of soil science based on the classification of soil type with climate and other formative factors developed in Russia, but the results were slow to reach researchers in Europe and North America.

Meanwhile an industry was born. Laboratory analysis that reduced plant nutrition to simplistic replenishment formulas using mineral forms of nitrogen (N), phosphorus (P), and potassium (K) were instituted. Universities set up agricultural departments and experiment stations that single-mindedly promoted their use. The huge manufacturing plants built to produce nitrogen for explosives in World War I were switched over to peacetime use making nitrogen fertilizers. The industrial-agriculture paradigm with its NPK mentality had gained ascendancy. The definition of soil, and how it was to be treated, was usurped by the laboratory and the fertilizer manufacturer.

The Dust Bowl

The rude awakening brought about by the Dust Bowl days in the 1930s led to a national effort to get agriculture back on a secure soil-fertility footing. Then secretary of agriculture Henry Wallace wrote:

THE EARTH is the mother of us all—plants, animals, and men. The phosphorus and calcium of the earth build our skeletons and nervous systems. Everything else our bodies need except air and sun comes from the earth. Nature treats the earth kindly. Man treats her harshly. He overplows the cropland, overgrazes the pastureland, and overcuts the timberland. He destroys millions of acres completely. He pours fertility year after year into the cities which in turn pour what they do not use down the sewers into the rivers and the ocean. . . .

This terribly destructive process . . . is not excusable in the United States in the year 1938. We know what can be done and we are beginning to do it . . . the public is waking up, and just in time. In another 30 years it might have been too late.

. . . No man has the right to destroy soil even if he does own it in fee simple. The soil requires a duty of man which we have been slow to recognize.[1]

The book from which this quote comes is *Soils and Men*, the 1938 Yearbook of Agriculture, published annually by the U.S. Department of Agriculture—the likes of which have not been seen before or since in USDA publications. In fact, whole sections of the 1,200-page tome read like an organic farming manual with titles such as "Loss of Soil Organic Matter and Its Restoration," "The Use of Cover and Green Manure Crops," "Crop Rotation," "The Nature and Use of Organic Amendments," "The Soil Requirements of Economic Plants," as well as a major portion devoted to the "Fundamentals of Soil Science."

Causes for the rampant soil degradation and widespread erosion of the Dust Bowl days outlined by *Soils and Men* were many and varied. Appraisals revealed that fully 61 percent of the country's cultivated soils had been badly misused and were partially or completely destroyed or had lost most of their fertility. Absentee ownership of the land, coupled with tenant farming, had

Without cover crop or mulch, tilled soil is vulnerable to wind and rain.

almost doubled to 42 percent over the previous fifty years, turning upside down the Jeffersonian ideal of a strong nation of independent yeoman farmer-citizens working their own place, stewards of the land and of democracy.

The protracted period of disastrously low farm prices in the 1920s and 1930s forced farmers to further overwork the soil, leading to destructive cropping systems to make the land pay, especially in commodities such as cotton, corn, wheat, and tobacco. The landlords and tenants alike needed to wring out every cent possible, and in the survivalist farm economy, there was no will for soil conservation measures. The Northeast was somewhat spared from the ensuing large-scale soil erosion problems because of the nature of pasture-based dairy farming and a much higher incidence of farmer owner-ship, where good soil practices were a solid investment in your family's future.

Although there has been some real progress in controlling soil erosion, a comprehensive report shows our national efforts are still unsustainable, despite the introduction of high-residue and no-till conservation tillage practices. Since 1982 erosion has been reduced about 38 percent, but that reduction has leveled off since 1995 and still amounts to losses of some two billion tons of topsoil per year. Fully 30 percent of our nation's cropland is eroding excessively. The upshot is the United States is losing soil ten times faster than the replenishment rate, with an economic impact estimated at $400 billion per year. Unlike the dire, in-your-face realities of the Dust Bowl, many of today's deleterious agricultural practices go largely unno-ticed, mostly hidden from view. While swirling clouds of soil no longer block the sun and impinge upon our consciousness, satellite photos reveal massive silt buildups in our heartland's major river systems and into the Gulf of Mexico. With the two-feet-deep prairie topsoils of the Midwest now down to four inches, the drama of erosion is more disguised and insidious as it quietly continues to gnaw away at the bone—the soil's skel-etal substance of humus and tilth, the essence of soil fertility itself.

Organic Fertility

In a career spanning the first half of the twentieth century, agricultural researcher Sir Albert Howard, considered the father of organic agricul-ture, significantly advanced the understanding of soil fertility as the basis for agricultural production. His work in Britain, India, and the West Indies produced important studies on the role of mycorrhizal fungal soil associations with plant roots, and he systematically developed composting methods to maximize microbial functions. Above all, he provided a new holistic basis to the scientific approach. His book *An Agricultural Testament* is an organic classic:

> The nature of soil fertility can only be understood if it is considered in relation to nature's round. In this study we must at the outset emancipate ourselves from the conventional approach to agricultural problems by means of the separate sciences and above all from the statistical consideration of the evidence afforded by the ordinary field experiment. Instead of breaking up the subject into fragments and studying agriculture in the piecemeal fashion by the analytical methods of science, appropriate only to the discovery of new facts, we must develop a synthetic approach and look at the wheel of life as one great subject and not as if it were a patchwork of unrelated things.
>
> All the phases of the life cycle are closely connected; all are integral to Nature's activity; all are equally important; none can be omitted. We have therefore to study soil fertility in relation to a natural working system and to adopt methods of investigation in strict relation to such a subject. We need not strive after quantitative results: the qualitative will often serve. We must look at soil fertility as we would study a business where the profit and loss account must be taken along with the balance-sheet, the standing of the concern, and the method of management. It is the "altogetherness" which matters in business, not some particular transaction or the profit or loss of a particular year.[2]

Scientific Reductionism

Today's conventional farmers might well ask why they need fertile soil anymore. Most extension service information is based on the use of soluble chemicals. Cheap fertilizers are readily available, and the application rates are the major part of the printout from the land grant college (and most other) soil-testing services. The Farm Credit banks base their operating loans on chemical input compliance. Formulaic farming has reduced the definition of soil fertility to its chemical constituents, highly oversimplifying the process. Crop-production efficiency becomes a narrow equation between synthetic inputs and crop yield, while the uncounted toxic side effects are externalized to the environment and public health.

Indeed, one of the hallmarks of industrialized agriculture is that it is highly reductionist. The problems occur when the truths discovered on a micro level in the lab are ratcheted up as absolutes into full-blown field applications that, in reality, exist in a higher, ecological realm. A few years back, for example, scientists delighted in depicting how human beings could be reduced to a precise list of chemicals with a shelf value of $5 or so. While they may be 100 percent correct in their analysis, it obviously doesn't go very far. No one is suggesting you can purchase those same chemicals, mix them back together, and come up with a human being— complete with opposable thumbs and animating life force. Yet this pattern of dismemberment is the "scientific" basis behind much of the technology that produces our food today.

Committing Biocide

It is important to realize the disastrous effects commercial chemical fertilizers have on soil life. Quite simply, most of them are toxic biocides—they oxidize, sterilize, disintegrate, and otherwise obliterate microorganisms and soil animals, on up through the soil foodweb. While synthetic fertilizers are designed to deliver soluble chemicals directly to the crop plants, bypassing the nutrient conversion processes (mineralization) by the soil, the effects of those chemicals are biologically disastrous. Chlorine is a common binder in many fertilizer compounds, for instance. It is also a potent bactericide, hence its use in bleach cleansers and swimming pool purifiers. Anhydrous ammonia, a nitrogen source increasingly selected for "safe" use on crop fields near waterways because it is not apt to migrate far, was used in World War II in the Pacific to help construct runways for aircraft on jungle islands. The liquefied gas is highly toxic to soil organisms, and repeated use compacts the soil to a cementlike consistency.

Once farmers start down the slippery slope of using chemical inputs, they have to continue—the biocidal effects on natural fertility throw the entire soil ecosystem out of balance and, over time, consume its contents. The truth is, most of the major croplands of the world are 100 percent dependent on agrochemical inputs. Literally, no chemicals = no crop. Their use continues to be justified by increased crop production—but the economic analysis fails to count the concomitant environmental and health costs or consider our dependence on petrochemical inputs.

Agricultural companies have long profited from this oversimplification of crop production. In a legal variety of a money-laundering scheme, a large percentage of government subsidy payments to farmers goes directly to paying the bills—for machinery, fuel, fertilizer, pesticides, and other inputs needed to produce the crop—and quickly ends up in the corporate coffers. There's also a huge industry of scientists, salesmen, academics, and system deliverers making their own livings in the background: a considerable vested interest in keeping things agribusiness as usual. The result is that the survivability of local, small-scale agriculture as well as the finite soil-fertility basis of our food supply is being silently and steadily consumed on a scale that makes the ravages of the Dust Bowl look almost benign in comparison.

Redemption

Ironically perhaps, redemption of holistic organics is happening in the lab as major discoveries in biology, genetics, and other scientific disciplines reveal the profound interconnectedness of all life. Modern technologies utilizing electron microscopes and cellular DNA analysis are revealing the complex biological interrelationships in the "soil foodweb" that provide the actual basis of plant growth. As science progresses, the "higher truth" of the soil-plant ecosystem supersedes chemical/transgenic, reductionist thinking about agriculture, confirming the organic paradigm.

The ancient stream of farmers' intuitive and practical knowledge is merging with and being verified by this new, comprehensive view, and it has the power to revolutionize agriculture. It has been a long time coming. Albert Howard staked out its parameters thus in 1945:

> A revolution in farming and gardening is in progress all over the world. If I were asked to sum up in a few words the basis of this movement and the general results that are being obtained, I should reply that a fertile soil is the foundation of healthy crops, healthy live stock, and last but not least healthy human beings. By a fertile soil is meant that Nature's law of return has been faithfully applied, so that it contains an adequate amount of freshly prepared humus made in the form of compost from both vegetable and animal wastes.
>
> This revolution in crop production involves making the very most of the earth's green carpet—that marvelous machinery for producing all our food and a great deal of the raw materials needed by our factories. Both units of this natural factory— the green cells of the leaf and the power which drives them (the energy of sunlight—owe nothing to mankind. They are the gifts of Providence which all the resources of Science cannot copy, still less improve. Mankind can only assist the food factory in two directions. He can look after the soil on which the green carpet rests and in which the roots of crops and the unpaid labour force of the soil—moulds, microbes, earthworms, and so forth—live and work. He can also by selecting crops and by plant breeding methods make the most of the energy of sunlight and of the improved soil conditions. But the plant breeder must avoid one obvious blunder. He must not be content with improving the variety only, otherwise his labours will soon lead to the exhaustion of the soil. The improved variety will take more out of the ground and will soon become a boomerang. The plant breeder, therefore, must always be careful not to confine his attention to the variety, but must increase the fertility of the soil at the same time. Such crops will look after themselves, and insect and fungous pests will do little or no damage.[3]

The Soil Habitat

A number of years ago in the middle of the growing season I was prepping a bed to plant the weekly succession of lettuce transplants when a grower from the farmers' market stopped by. I was in the process of spreading soybean meal, a high-protein animal feed, on the newly tilled beds in preparation for rotovating it under. My visitor was aghast at my "wasting" good feed and asked if I had been out in the sun too long. When I explained that I was feeding my soil animals, he was convinced I had lost my mind. Telling him that soybean meal—dried beans processed into golden flakes—was a wonderful organic soil amendment—whose 47 percent protein label on the bag translated into 6 percent nitrogen after being digested by soil microorganisms (just right for lettuce)—didn't help much either. He could see feeding it to livestock like you're supposed to and then spreading the manure, but dumping it on the ground was sinful. He couldn't argue about the high quality of produce I routinely brought to market, but plainly I had crossed the line into what in his mind were deviant practices.

Nowadays not only is soybean meal too expensive a nitrogen source owing to its high demand as a high-protein animal feed supplement, it is more apt to be processed from genetically engineered beans. It serves, however, to illustrate the connection between plant protein and nitrogen as well as the role soil microbes play in the process. It also points out the difference between bioavailable soil amendments and soluble chemical fertilizers.

Once tilled in, the soybean meal has to first undergo a process of bacterial breakdown and digestion by soil organisms. To tackle the job, the decomposer bacteria greatly multiply their numbers (already at many millions per teaspoonful in a fertile soil) as they consume the feed. The soy protein is thus converted to microbe protein in their bodies. As their greatly expanded numbers fall prey to protozoa and other soil animals, their body protein becomes food for diverse microbial specialists that liberate (or "mineralize") the previously bound-up organic compounds of nitrogen, phosphorus, and other nutrients, converting them into soluble forms that are usable by plants. Some of the details of these processes are discussed in chapter 3. In a process that is key to organic management, nutrients are stabilized and retained in the plant residues and microbial biomass, prevented from leaching and running off or volatilizing into the air.

Room and Board

The organic grower's primary task is to provide high-quality room and board for the legions of microbes and soil animals that do the actual work of growing the crops. Organic matter does both—not only feeding the soil biota but also providing them a beneficial habitat for settlement and proliferation.

For microbes and plant roots alike, the physical "room," or soil tilth, is just as important as its nutrient, or "board," content. Poor drainage, for instance, can limit even the most fertile soil. Soil textures are physically categorized by their particle size—from sand (2.0 to 0.05 millimeters) to silt (0.05 to 0.002 millimeter) to clay (< 0.002 millimeter). Loams have moderate mixtures of all three.[1] While your soil may range from a clay loam to a loamy sand, its tilth describes its workability (ease of tillage) and hospitality to microbes as well as to seed emergence and root penetration. Further classification of wet or droughty may correspond to clay or sand content or to the overall geological conditions of the area.

The single remedy that improves all soils—whether too wet, too droughty, too sandy, too heavy with clay, too compacted, and so on—is the regular incorporation of diverse organic matter. Over time, a buildup of organic matter ameliorates structural extremes and helps bring the soil ecosystem into balance. As it amplifies soil porosity, it reverses compaction from heavy equipment. Moisture-holding and drainage capacities are both promoted.

Most critically, soil air is increased. Soils need to breathe; microbial respiration (O_2 in, CO_2 out) requires plenty of oxygen. Roots, too, require porous soil as they seek the path of least resistance for their growth, and oxygen-rich soil water promotes their uptake of nutrients. Practices that prepare a good root bed down under are just as important as providing a fine seedbed up top. Chisel plows and subsoilers, while requiring extra tractor horsepower to operate, help fracture and open up dense soil layers and hardpan to air and water. Cover crops such as cowpeas, sweet clover, and a sorghum–Sudan grass hybrid are able to punch their roots through the subsoil. (Many cover-crop seed varieties are sold laced with fungicides and other "crop-protection chemicals"—be sure to order untreated seed well ahead of time.) While the incorporated plant tissue biomass provides large quantities of organic matter for soil microbes, the decomposing roots open deeper passageways for air, water, and future crop root growth.

Solubility

The beauty of bioavailable soil amendments is that they are relatively stable in the soil, available at the plant's request—which may vary considerably from moment to moment, day to day. During cool and cloudy

Bioavailable soil amendments supply nutrients on demand. Chemical fertilizers force growth, which invites predation by pests.

weather, for example, plants may slow their metabolism, requiring significantly less nutrients for the period. When the sun comes out after a rain, their nutrient needs accelerate and they resume feeding. The immediate growth requirements of the plant are the deciding factor. Approximately 20 percent of the nutrient energy photosynthesized by the plant's leaves goes directly to feed and colonize the billions of beneficial microorganisms in the *rhizosphere* (root zone) through root exudates.

In this regard, cold soils suppress microbial as well as plant-growth activity. Although many soil organisms break dormancy when the soil thermometer registers 40°F (4°C) or so, they don't really become active enough to support key disease-suppressing or mineralization functions until temperatures stabilize above 60°F (16°C).

In stark contrast, soluble chemical fertilizers dose the plant directly, and the plant has no choice but to go into an unbalanced, rapid growth spurt no matter what its needs. Researchers have found that this produces fast, fleshy growth high in simple sugars and amino acids, which is highly attractive to pests.[2] The researchers also found that plants that have a good mineral balance from organic sources quickly produce complex starches and proteins that are not attractive to insect pests. These basic differences in plant-nutrition approaches come down to this: amendments feed the soil; fertilizers dose the plant.

Holding on to Humus

Not all forms of soil organic matter are readily decomposed by microbes. Complex cellulose compounds along with plant lignins, fats, and waxes, for example, are much more resistant to decay. A more lengthy biological process called *humification* may take place in soils where organic matter is further distilled and converted to more stable substances by microbial activity. As Selman Waksman wrote in *Humus*:

> The plant and animal residues do not become completely mineralized. A certain part of those residues is more or less resistant to microbial decomposition and remains for a period of time in an undecomposed or in a somewhat modified state, and may even

accumulate under certain conditions. This resistant material is dark brown to black in color and possesses certain physical and chemical properties; it is usually called HUMUS. As a result of the formation and accumulation of this humus, a part of the elements essential for organic life, especially carbon, nitrogen, phosphorus, sulphur, and potash, became locked up and removed from circulation. In view of the fact that the most important of these elements, namely, carbon, combined nitrogen, and available phosphorus, are present in nature in only limited concentrations, their transformation into an unavailable state, in the form of humus, tends to serve as a check upon plant life. On the other hand, since humus can undergo slow decomposition under favorable conditions, it tends to supply a slow but continuous stream of the elements essential for new plant synthesis.

Humus thus serves as a reserve and a stabilizer for organic life on this planet. It also plays a prominent, if not predominant, role in the formation of most soils. It exerts a variety of physical, chemical, and biochemical influences upon the soil, making the soil a favorable substrate for plant growth.[3]

In terms of furnishing first-rate room and board accommodations for soil life, humus conditions the soil. It creates a well-aggregated habitat as well as supplying an immense reservoir of nutrients for microbes and plants alike. The stable humus content of the soils we work today may easily be hundreds or thousands of years old. While the process of humification is ongoing, it's exceedingly difficult to increase this form of organic matter in active agricultural production systems. Misguided farming methods, however, can readily affect and diminish this core resource. Best management practices can only hope to preserve what's already there.

Primarily this means regularly incorporating organic matter into the soil as a buffer to protect humus—the more diverse the source, the better. Rotating crops varies the quality of the crop residues that are cycled back into the soil, with legumes being number one on the list. Composted manures boost bacterial populations, while rotted leaves favor the fungal side of the spectrum. Both are required to provide healthy crop nourishment.

Cover crops have their own humic qualities. When a lush buckwheat cover crop is turned into the soil, for example, the decomposer microbes go to work breaking down and digesting the plant materials. While the fleshy green leafy materials are easily consumed and quickly become available nutrients for other soil animals and plants, they do not form humus but are an immediate fertility booster. Immature, fleshy cover crops for this purpose are called green manures because, like animal manures, they are high in immediately available nutrients.

Left to grow longer in the field, mature cover crops build up woodier stems and more carbonaceous materials containing mixtures of complex compounds that take more time and microbial energy to break down and digest. This is known as active humus—dark, crumbly, sweet-smelling soil organic matter whose remnants of stems, leaves, and other components are still somewhat recognizable by eye. It provides a longer-term, steady release of nutrients as the more specialized decomposer bacteria continue the digestion process, and it helps to buffer the soil ecosystem from the untoward effects of everything from acid rain to protracted drought.

Stable humus, on the other hand, is the highly decay-resistant material that remains after the complex digestion processes have run their course.

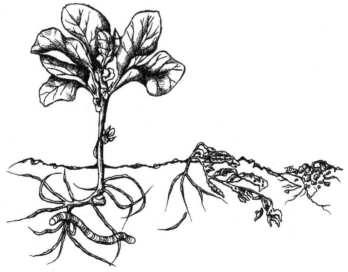

A cover crop decomposing–leaf, stem, and root–into nutritious, humus-protecting organic matter.

It can only be seen under a microscope. While no longer rich in available nutrients, it is the organic essence that forms the structural backbone of the soil, helping to bind it together and providing essential habitat for soil life and root growth, as well as promoting good drainage and moisture-holding capacity.

Soil to Burn

While good organic soil-management methods such as extensive cover cropping, spreading of compost, and rotation in and out of sod crops can build up active humus on the one hand, it is important not to reduce them by excessive injurious cultivation practices on the other. Farmers and gardeners should recognize that plowing, rotovating, power spading, and cultivating, while all having their purpose in forming a workable seedbed, also stimulate decomposition and oxidize soil assets. Minimizing tillage maximizes soil fertility.

Anyone familiar with wood heat knows the importance of controlling the air intake to promote a long, steady release of heat energy—to keep things warm and also conserve the woodpile. In the book *Building Soils for Better Crops*, coauthor Fred Magdoff likens the decomposition of organic matter in the soil to burning wood. When wood burns,

> the carbon in the wood combines with oxygen from the air and forms carbon dioxide. As this occurs, the energy stored in the carbon-containing chemicals in the wood is released as heat in a process called *oxidation*. The biological world, including humans, animals, and microorganisms, also makes use of energy inside carbon-containing molecules. This process of converting sugars, starches, and other compounds into a directly usable form of energy is also a type of oxidation. We usually call it respiration. Oxygen is used and carbon dioxide and heat are given off in the process.[4]

Such respiration can be too rapid. "The more a soil is disturbed by tillage practices," Magdoff writes, "the greater the potential breakdown of organic matter by soil organisms," and

Rapid decomposition by soil organisms usually occurs when a soil is worked with a moldboard plow. Incorporating residues, breaking aggregates open, and fluffing up the soil allow microorganisms to work more rapidly. It's something like opening up the air intake on a wood stove. . . . In Vermont we found a 20 percent decrease in organic matter after five years of growing corn on a clay soil that had previously been in sod for a long time. In the Midwest, cultivation of soils for about 40 years has caused a 50 percent decline in soil organic matter.[5]

This analogy takes a literal turn in John McPhee's narrative describing the highly organic muckland (as opposed to mineral) soils in Pine Island, New York:

The soil is so rich it will burn. Now and then a farmer flicks a cigarette off his rig and starts a ground fire. Old John, who smokes, has put out three such fires on his fifty acres this summer. Blazing black dirt is not easily controlled. In 1964, there was a fire in which a couple of thousand acres burned. The fire crossed many property lines. It was harvest time, and crated onions were stacked across the plain. Fifteen million onions roasted. The fallout quickened appetites in Greece. "There were big winds. You couldn't go nowheres evenings. The air was too thick to see." A tractor blew up. The fire engines of many towns and a firefighting helicopter converged upon the muckland without significant effect. The fire lasted more than a week, and went out under heavy rains.[6]

Most soil-testing services offer an organic matter (OM) analysis given as a percentage, but it sometimes has to be specifically requested. Labs may also use different testing methods that will yield different values. By the wet chemistry method, for example, a good soil might register a value of 3 percent organic matter. Using a "weight loss" measurement at high temperature, the same soil may register 4 to 5 percent.

Even in a well-developed loam, however, the percentage of organic matter rarely is more than 5 percent, while mineral components might make up 45 percent and water and air check in with 25 percent each. That same loam with an organic matter content of only 2 percent, however, would be on its last legs in terms of tilth and fertility, while a sandy soil with 2 percent OM might be relatively healthy.

Hard numbers aside, most important for growers is the relative increase or decrease in the OM values over time. Downward trends may indicate excessive tillage practices, for example, while increases may help verify that certain rotational or cover-crop methods are working.

two

Sustaining Soil Fertility

The simple definition of a fertile soil has to do with its capacity to nourish healthy plants. From a holistic viewpoint, however, this is far from a simple equation. Soil fertility is a dynamic state in which intricate, balanced, healthy relationships among soil structure, soil biota, and nutrients work synergistically in a complex ecosystem. Fertility is an energetic condition: soil is alive and requires sustenance and space for aeration. When healthy, it effectively performs its functions. It feeds crops of itself and requires replenishment with organic matter. Fertile soil has the capacity to resist and suppress pests and diseases. It retains moisture in droughty times and drains well when conditions are wet. It sequesters carbon, nitrogen, and other nutrients that otherwise pollute our water and contaminate the atmosphere. Ultimately, our lives depend on it.

Soil fertility is regarded more as a quaint, outmoded concept by modern agriculture. Under the industrial criteria of quantifiable inputs and production yields, fertilizers and pesticides reign supreme. However, "high-yield agriculture" conveniently forgets to count the non-targeted side effects of these inputs that routinely degrade ecosystems and pollute the environment. The substantial manufacturing, transportation, and application costs also subtract from the net yield.

Fundamentally, production agriculture has substituted oil for soil to achieve its yields. With world oil productions expected to peak in the coming decade, we are entering a time of increasing petroleum scarcity and market instability. Natural gas, the most common hydrogen source for manufacturing nitrogen fertilizers, is also in limited supply and increasingly expensive. Drilling in fragile wilderness areas is becoming more unacceptable.

Sustainable organic systems, in contrast, often utilize legume cover crops inoculated with nitrogen-fixing bacteria to capture atmospheric nitrogen in the root nodules. Not only does this supply a stable, biologically available amount of nitrogen when the plants are turned in to feed the soil for the following crops, but the organic residues benefit the soil.

Testing, 1, 2, 3, . . .

Most land grant university soil tests can be an exercise in frustration for organic farmers. Nutrient values are usually given in pounds per acre of synthetic chemical inputs, and translating them into natural equivalents is difficult. Fertility is defined mostly in terms of the available levels of macronutrients that crops need in the largest amounts for optimum growth. Nitrogen, phosphorus, and potassium (N, P, K) are the "big three" nutrients listed on a bag of 10-10-10 fertilizer. Calcium, magnesium, and sulfur complete the standard soil test's list of nutrients. Although such simplicity is valuable in identifying gross nutritional imbalances (under Liebig's law the nutrient in greatest deficiency must first be supplied before any other added elements can improve crop production), it ignores the host of micronutrients (also called "trace elements," including iron, manganese, zinc, copper, boron, and molybdenum) that plants require in minute quantities.

In the long run, chemical fertilizers are quite toxic to soil life and can easily end up polluting the environment through leaching and runoff. Sustainable soil-fertility practices, on the other hand, build stable nutrient levels in the soil, which in turn provide for the needs of the crop. While there are a number of identical physical, chemical, and biological functions that must be performed in each system, losing sight of the soil's aliveness has further dire consequences to ecosystem health.

Although standard soil tests give most values in terms of soluble chemical nutrients rather than vital fertility factors, there are soil-health assessment guides that help organic growers self-evaluate the specific qualities of their own farm or garden. Ohio State University Extension, for example, has a short "Soil Health Card" to facilitate the process.[1] Checklists to assess soil tilth factors such as crumb structure, crusting, and compaction are enumerated. Soil-life indicators including clean earthy smell, earthworms, and residue decomposition are presented, and evaluations of drainage and moisture-holding capacity enter into the overall analysis.

Likewise, the book *Building Soils for Better Crops* presents a comprehensive table of visible "qualitative soil health indicators."[2] Earthworms as an indicator, for example, are best assessed in either the spring or the fall when there is good soil moisture. A poor soil is defined as having 0–1 worms per shovelful of topsoil, with no casts or holes present. A medium soil has 2–10 worms, and a good soil has more than 10 worms with lots of casts and holes as well as birds following behind tillage. Another factor, subsurface compaction, is measured by the ability to insert a wire flag down into the subsoil.

Another good overall resource on soil health is ATTRA (Appropriate Technology Transfer for Rural Areas), the sustainable agriculture information center funded through USDA. In addition to providing free assistance to farmers, along with publications and other resources, it offers free, updated listings of alternative soil-testing laboratories.[3]

Finally, the Sustainable Farming Connection Web site[4] offers a comprehensive listing of materials and links to resources regarding soil health and many other aspects of alternative agriculture.

May the Force Be with You

Soil is subject to the same electromagnetic force that holds all materials—gaseous, liquid, or solid—together. Positive and negative charges attract one another; like charges repel. While some of the resultant combinations (compounds) fulfill each other's charges completely, many others—even while already bonded in compounds—still have an excess charge. These positively or negatively charged molecules are ionic, capable of further attraction and bonding.

Colloid (from the Greek word *koll*, "glue," and *oid*, "like"), refers to the electromagnetic ability of minuscule soil particles to latch on to nutrients. Clay and humus particles, for example, are *anions*, possessing a powerful negative charge able to attract and bind positively charged essential nutrient *cations*, such as calcium (Ca), hydrogen (H), potassium (K), and magnesium (Mg). Major soil anions are nitrogen (N), sulfur (S), aluminum (Al), and phosphorus (P). Their molecules—nitrate (NO_3), sulfate (SO_2), and phosphate (PO_4—are also anions, ready to bond to cations such as Ca, H, or K. Finally, some ions, including carbon (C) and silicon (Si), can go both ways and bond with either cations or anions.

An important measurement of a soil's colloidal reservoir, therefore, is its cation-exchange capacity (CEC). On a conventional soil-analysis report, major nutrient levels are given in pounds per acre that are available for uptake by the plant. Actual availability, however, has to do with the soil's ability to both hold and release nutrients, and this is what the CEC measures.

Because the CEC of a soil is mostly determined by its humus and clay content (the primary colloids), clay soils and soils with a large organic matter content will rate high on the 0–15 scale, with the ability to hold and deliver ample amounts of nutrients. A sandy or gravelly soil with low organic matter will have a low CEC (less than 7) and a considerably diminished capability to bind or provide nutrients. CEC is also a useful measurement for determining moisture-holding capacity, because soil colloids have a greater capability for attracting and retaining water. Not surprisingly, a sandy or gravelly soil with low organic matter isn't going to be able to hold much moisture either.

Trace Elements and Rock Dust

Organic molecules are large, complex (and not completely understood), latticelike structures able to hold on to large amounts of nutrients until they are needed by plants. They are also able to isolate toxic elements and prevent them from doing harm. *Chelation* (from the Greek word *khele*, "claw") refers to the molecule's capacity to utilize more than one bond for tightly gripping single nutrient ions. Because plants require trace elements in very small quantities, and imbalances or overdoses can be toxic, chelates can sequester excess micronutrients until they are needed.

Raw-mineral soil amendments, such as greensand, rock phosphate, and granite dust, generally provide ample trace elements in correctly balanced amounts, but they need to be mineralized by soil microbes to become available to plants. While the trace-element content of organic matter totally depends upon the mineral resources of the soils it grew in, the more diverse the materials that are added to a soil, the more plentiful it will be in chelates, and the greater nutrient-holding capacity it will possess. Because plants are extremely sensitive to excess amounts of trace elements, adding more for fertilizer should only be done in accordance with strict soil-test recommendations to prevent toxicity.

The allure of a simplistic "magic bullet" approach, so prevalent in chemical farming, has counterparts in alternative agriculture as well. The "soil remineralization" movement is a case in point. Based on the premises of John Hamaker in the 1970s, coauthor with Don Weaver of *The Survival of Civilization*, rock-dust proponents maintain that finely ground glacial gravel dust (along with volcanic ash) is nature's way of creating fertile soils and is the necessary missing ingredient to replenish farmland and increase crop yields along with nutrient levels in our food supply. Their advocacy of blanket applications (at a rate of 10 tons per acre) of untested local rock fines from the stone industry, gravel pits, cement operations, and the like, can be a dangerous practice.

At the very least, this approach puts the cart before the horse. While there is no question that granite, basalt, and other rock powders may be rich in both macronutrients and trace elements, they are useless—or toxic even—without ample amounts of soil organic matter to hold and buffer them, a balanced pH to make them available, and enough room and board for the mineralizing microorganisms. Adding copious amounts of unanalyzed rock dust to infertile soils with little tilth and low organic matter could be a recipe for disaster. Needless to say, mining, grinding, and then shipping high-analysis rock dusts across the country has further environmental and economic effects.

It's not that there aren't other nutrient sources naturally available. Many soils have a seemingly unending supply of stones that are constantly in the process of becoming soil, breaking down through the action of plant and soil acids and weathering. Each pass with a steel plow or cultivator accelerates the process.

Precipitation regularly carries atmospheric inorganic nitrogen compounds, such as ammonium and nitrate, into the soil. A late spring snowfall is known as "poor man's manure" for its ability to green up fields and lawns alike. Even industrial pollution supplies increased amounts of beneficial airborne nutrients such as sulfur and nitrogen, the acidic effects on plant ecosystems notwithstanding.

Finally, *Azotobacter*, a genus of soil bacteria that thrives in soils with high organic matter, has the ability to fix nitrogen directly from the air. The differing rhizobia strains growers routinely use to inoculate legume seed before planting accomplish the same thing. To maximize nitrogen, however, legume cover crops should be turned under before they set seed, as the reproductive process utilizes much of the plant's nitrogen reserves.

Soil Limitations

Although a soil profile might look good on paper, values such as cation-exchange capacity don't always tell the whole story. A clay soil may have a high CEC, for example, but coupled with low organic matter may lack soil air, another critical element for microorganisms and plant roots alike. Such hostile conditions select for specific anaerobic ("without air") microorganisms, displacing their more beneficial counterparts. The lack of oxygen also retards the decomposition rate of organic matter. Over time, if enough organic matter builds up in the soil system, a shift back to conditions favoring aerobic microbes may occur, rebalancing the soil ecosystem. An over-abundance of clay colloids can also compete against crop roots for soil water and its dissolved mineral nutrients, starving the crop.

Clay soils are difficult to work. Timing tillage to proper soil conditions is critical. Plowing or cultivating when it is too wet collapses the structure and diminishes soil air. The crusting these soils are prone to can be a barrier to seed germination. Droughty conditions produce cracks and fissures that open up and expose soil depths to further drying.

Working with the forces of nature is the best strategy. While the general rule is to always keep the ground covered as much as possible, tilling clay soils in the fall on land where erosion isn't a threat leaves the soil open to freeze/thaw cycles, which break up clods and produce finer seedbed tilth. Such heavy soils are often unworkable during prolonged wet spring conditions, and a whole season may be lost waiting for the proper state for tillage.

The sandy-gravelly-type soils with low moisture capacity suffer from the opposite extreme. There is relatively too much air in the system, accelerating the oxidation of organic matter and further diminishing soil capabilities. Here, turning in copious quantities of compost and cover crops is the way to go. The beauty of ample soil organic matter, again, is that it buffers all soil extremes, holding moisture in dry times and promoting drainage when it is too wet.

While soils that are too sandy or have too much clay are not suited for farming to begin with, most soils have some intrinsic limitations. Extreme weather conditions of extended drought or protracted rains can stress the most balanced soil ecosystems. It is not economic to try and rebuild the soils themselves: adding truckloads of sand to clay soils, or clay to sandy soils, will have far less remedial effect than the concerted use of compost and green manures.

Besides the addition of plentiful amounts of organic matter, there are numerous strategies for overcoming certain soil problems. For poorly drained areas, tiling is a time-honored practice made easier and less expensive by the use of corrugated plastic drainage tubing and slit-trenching equipment. For soils with even a slight slope, berms (mounds) and swales (wide, shallow ditches) can direct the flow of excess water runoff out of the fields. Constructing ponds can sometimes utilize wet or poorly drained areas to best advantage. Wetland areas should be preserved, however. Their habitat value for beneficial species plays an important role in the farm ecosystem.

Overall, a soil should be worked within its limitations. The choice of soil amendments can either ameliorate or exacerbate unbalanced soil situations. A colloidal, soft rock-phosphate product with its high clay content would be a good choice for application on a loose, sandy soil, for example. Clay soils, on the other hand, would benefit by using a gritty-textured, hard rock-phosphate product.

Aggregation

The bonding capacity of the primary soil colloids—clay and humus molecules—depends upon their open surface area as well as their ionic charge. The more colloidal-exchange sites are available, the more nutrients the particular soil can hold. Further, the smaller the particles, the greater their combined surface area. Doing the math, if all the particles contained in the plow depth of an acre of soil had diameters of 1 millimeter each, the total interior surface area would be about 500 acres. The diameter of colloidal particles, however, is defined as 1 micron (0.001 millimeter) or less. If all the particles in that same acre zone were one micron each, the internal surface area would be 500,000 acres.[5]

The fertility of a soil is just as much a function of its structure as it is of the life it contains and its nutrient capacity. Humus has a complex aggregate (crumb) structure composed of variously sized and shaped particles extremely conducive to bonding nutrients. Bacteria produce slime layers that help bind soil particles loosely together. While clay has multitudinous small particles, they tend to stick together in flat, tightly overlapping platelets, yielding a minimal surface area. Once again, incorporating quantities of organic matter, through composts, green manures, and cover crops, dramatically develops the aggregation of clay-based soils, as well as increasing their cation-exchange capacity. Sandy soils have no definite structure and may contain little colloidal material, but they benefit immensely from added organic matter as well.

In addition to offering nutrient-exchange sites, a well-aggregated soil promotes aeration and drainage and creates a habitat for all shapes and sizes of soil life, from bacteria and fungi to centipedes and earthworms. It is also highly conducive to root growth. The rhizosphere (the area around each root hair) is a dynamic nutrient-exchange zone, pulsating with billions of microorganisms that mineralize nutrients for plant growth and also feed on photosynthesized nutrients delivered by the plant.

Maintaining a well-aggregated soil structure enhances soil fertility. This means never working soils when they are too wet. A squeezed handful of soil should not stick together, but fall apart afterward. The same goes for getting on the field with equipment too soon in the spring or after a heavy rain: the resulting compaction destroys soil structure, and that early head start can come back to haunt you for years to come.

hree

The Soil Foodweb

In our self-centered view of the world, we humans like to see ourselves perched atop a food-chain hierarchy. Plants, animals, and other creatures are here at our disposal; they're our food. Our intricate interdependency with them is conveniently ignored.

Rather, a holistic view of humankind holds true: we are life-forms joined with countless other life-forms in the much broader totality of an interconnected web of life. In this web, we are not the spider feeding supremely at the center but a coevolved species linked together with others in a woven and rewoven network.

New Tools, New View

A number of present-day technologies—from differential interference microscopy to computerized biological microanalysis and DNA testing—are giving researchers new tools to see deeper into the elaborate workings of soil ecosystems. Breaking new ground, Dr. Elaine Ingham, working now at Oregon State University, has put together a comprehensive and revolutionary view of soil/crop interaction that holds immense value for farmers, gardeners, and, indeed, all of world agriculture. Available on the Internet at www.soilfoodweb.com, her research into the intricacies of the "soil foodweb"—the complex relationships among soil microorganisms that govern nutrient mineralization, availability, and retention—has yielded qualitative and quantitative values that help define healthy soil ecosystems and sustainable crop productivity. She has also established a database of microbial soil samples from around the world to help determine what food-web system can enable farmers and gardeners to raise crops without pesticide and fertilizer dependency.

Her work serves to put Justis von Liebig's reductionism finally to rest. The fact that plants can live on strictly soluble chemical diets (as with soilless hydroponic systems, for example) tells only a small part of the story. The continuously required doses of "crop-protection chemicals" necessary to produce a crop this way smacks of a patient on life support and is far from a state of health. Soil food-web analysis, on the other hand, makes use of actual microscopic counts of the variety of decomposer organisms and their predators as well as their relative biomass, activity, and community structure. Because of the critical functions the microbes perform, their species populations and diversity are prime indicators of a healthy soil/crop ecosystem.

At the heart of Dr. Ingham's work is a much greater understanding of what is occurring in the rhizosphere—the dynamic interface of soil and root—and therefore, how nutrients are made available and how crops feed. Fully 60 percent of a row crop plant's solar-generated energy goes directly to the root system; for trees it's 80 percent. Once the work of growing structural and lateral root networks is established, fully half of the plant's root-directed energy goes into producing root exudates. Dr. Ingham refers to the material of these exudates as "cake"—a mixture of simple sugars, carbohydrates, and proteins manufactured by the plant to feed and breed up populations of compatible microorganisms. What's more, individual varieties and species of plants manufacture different, specific kinds of cake designed to attract and colonize particular beneficial microbes in their rhizosphere. In their growth cycle from seedling to maturity, plants also create different versions of their recipe to attract the microbes appropriate to that stage of growth. So singular is the plant/microbe relationship that microanalysis of a soil sample alone can identify the particular plant cultivar that was growing in that soil as well as the time of year the sample was taken.

Colonization

In essence, each plant is encouraging the kind of beneficial microorganisms that will breed up and colonize the surrounding soil and, in turn, seek out, make available, and deliver the needed nutrients back to the plant. A single plant, therefore, may be supporting billions of bacteria to provide its own needs. The microbes also protect the surrounding soil and root sites from invasion by disease organisms. This is a hugely vast

Root system and microscopic view of life around the root.

and complex system. Each gram (a teaspoonful) of fertile soil from the biofilm surrounding the crop root hairs is teeming with some 600 million bacteria. They are the primary decomposers, adept at multiplying their numbers to convert the organic matter nutrients into their own bodily biomass and immobilize them in the soil.

The next stage, the mineralization of this nutrient-retaining bacterial biomass into plant food, is accomplished by the balanced interactions of legions of other microbes and larger soil animals present in a fertile soil. The predators of the bacteria—protozoa, nematodes, microarthropods, and earthworms—in turn become prey for millipedes, centipedes, beetles, spiders, and voles. Through these many interactions the soil foodweb makes nutrients available to plants for their immediate growth needs as well as retaining additional nutrients for future crop demands.

Pipelines

Soil fungi also decompose organic matter, possessing a reach extending well beyond the rhizosphere. They search out, mineralize, and deliver soil-borne nutrients directly to the plant roots. One family of beneficial fungi, mycorrhizae, sends long strands (hyphae) throughout the immediate soil area and generates organic acids to dissolve mineral nutrients directly from the surfaces of stones and rocks, delivering them back to the plant's root system. The hyphae also act as pipelines reaching down to lower water table levels and nutrients in the subsoil, pumping water and supplies of nitrogen, phosphorus, boron, iron, and the like back up to the plant. In a droughty year particularly, this augmented water supply can make or break a crop.

The balance among soil organisms is critical. Alterations in the bacterial-to-fungal biomass ratio alone, for instance, can severely affect the structure and viability of plant communities aboveground. While it has long been known that chemical fertilizers and pesticides can severely affect both the diversity of soil species and their numbers, researchers can now measure ecosystem degradation directly by measuring the active and total biomass of each microbe group:

> plants grown in soil where competing organisms have been knocked back with chemicals are more susceptible to disease-causing organisms. If the numbers of bacteria, fungi, protozoa, nematodes and arthropods are lower than they should be for a particular soil type, the soil's "digestive system" doesn't work properly. Decomposition will be low, nutrients will not be retained in the soil, and will not be cycled properly. Ultimately, nutrients will be lost through the groundwater or through erosion because organisms aren't present to hold the soil together.[1]

Bacterial/Fungal Balances

Even under organic management, many farm and garden soils tend to have low mycorrhizae counts. A 1:1 ratio of bacteria to fungi is ideal for most cropping systems. Micro soil analysis has revealed, however, that row-cropping systems slip over time toward predominantly bacterial soil associations, while forest and orchard soils are fungus dominated. Bacterial/fungal balance not only has vast implications for plant nutrition and subsequent food quality (including higher protein and more flavor); it also expands the basic definition of soil fertility to include healthy balances of beneficial microbes. Dr. Ingham has established a soil-testing service (also accessible through www.soilfoodweb.com) that provides a direct biomass analysis of soil samples with actual counts and measures of the constituent microorganisms to help assess these factors.

Standard agricultural practices—frequent tillage especially—damage the hyphae of the mycorrhizae, breaking up the pipelines. Periodically rotating fields to sod crops is an excellent way to restore fungi to a soil. Methods that utilize mulches or leave crop litter on top of the soil also help select for fungi. So do composts made with tree bark and other highly carbonaceous materials.

More holistic system approaches include strip tillage where mycorrhizae-encouraging sod crops are grown side by side with row crops. Strip insectary intercropping, or biostrips (see chapter 7) are a further refinement of this practice, integrating pathways of perennial herbs, grasses, and wildflowers to provide a full-time fungal habitat alongside the raised beds. After the beds are tilled and planted, the mycorrhizae that live in the biostrip pathways can quickly reinvade the adjacent raised-bed areas and colonize the crop plants. Mycorrhizae also produce spores that can remain dormant until suitable conditions prevail in the soil environment.

Doing the Mineralization Math

Another major contribution of soil food-web analysis is a growing understanding of how soil organisms mineralize nutrients for plants from organic matter. In the vast microbe-eat-microbe soil world the larger creatures—various protozoa (including amoebae), beneficial nematodes (not the parasitic root-feeding species), and arthropods—feed on the huge populations of bacteria and fungi that are busy breaking down and digesting soil organic matter. Because the higher soil animals require less nitrogen-rich foods than their prey bacteria or fungi, the excess nitrogen is released in the process in the form of ammonium (NH_4). This oxidizes in the soil with the help of aerobic bacteria, into nitrate (NO_3^-), the most efficient form of nitrogen for plant growth.

This occurs via the various microbes' carbon/nitrogen (C/N) ratios. Bacteria have five carbons to one nitrogen, while the larger protozoa have 30:1. (Incidently, humans also weigh in with C/N ratios around 30:1, although Dr. Ingham explains that some people who have put on additional girth over the years may be approaching ratios of 40:1 or more.) Under this equation, a hungry protozoan must consume six bacteria (containing five carbons each) in order to satisfy its carbon requirement of 30. The protozoan only needs one nitrogen, however, liberating the other five "nitrogens" into the soil as freed nutrients available to plant growth.

Beneficial nematode species, feeding on diverse bacteria, fungi, algae, yeasts, diatoms, and other nematodes, need even less nitrogen than do protozoa. With C/N ratios near 100:1, a nematode must eat 20 bacteria to meet its carbon requirements and after consuming one nitrogen, releases the 19 others into the rhizosphere for use by plants.

Soil Degradation

When soil ecosystems are mishandled or thrown into imbalance by adverse environmental impacts, problems can show up quickly. Rapidly increasing numbers of bacteria, for instance, can tie up nutrients in their bodies, limiting the nitrogen available for plant growth. Temperature increases due to global warming and climate change could produce this effect. Alternatively, increases in global carbon dioxide (the "greenhouse effect") can inhibit bacteria, affecting the decomposition rate of organic matter and thus affecting nutrient cycling and soil structure. The tendency of the soil ecosystem to retain and recycle nutrients through microbial decomposition and mineralization helps prevent "nonpoint-source" runoff—pollution endemic to the use of soluble chemical fertilizers.

A balanced and healthy soil foodweb also creates a well-aggregated, friable soil structure that forms a beneficial habitat for microbes. The bacterial slime layers and threads of fungal hyphae bind soil particles against erosion and provide spaces for the oxygen needed by beneficial aerobic organisms. Ingham says healthy soils will spring back after compaction owing to the heightened biological activity. On the other hand, degraded soils will further shut down functions as the aerobic microbes "go to sleep" and the virulent anaerobic organisms come to the fore. The toxic compounds and alcohol they produce kill roots. The nose knows them as the smells of sour milk (butyric acid), vinegar (acetic acid), and vomit (valeric acid), and their presence turns the soil to hardpan.

Because imbalances in the soil will eventually manifest as ill health in crops, natural vegetation, and on to livestock and humans, it pays to rectify problems at the base level before they expand and get out of hand. The basic remedy for any soil is to regularly add compost, cover crops, and other forms of organic matter. The soil foodweb's capacity to resist and suppress disease organisms is highly complex. Diverse numbers of species are needed to combat them. Ingham says that "bugs in a jug"—commercial microbial preparations—contain only a limited number of species, not enough to do the job.

Diversity in the foodweb also produces more nutritious fruits and vegetables. Dr. Ingham cites the example of grapes growing in a balanced soil foodweb that had three times the protein of those growing in a depleted system; wheat had ten times more, while organic produce had higher protein contents across the board. Nutrient uptake is particularly increased when plants are grown with abundant mycorrhizae.

Ag Implications

Soil food-web research underscores the prime organic directive, "Feed the soil, not the plant." The microscopic world of soil organisms has been opened for all to see and can no longer be ignored. The microbes' incredible numbers and diversity, along with the intricate complexity of their interrelationships, gives a new, more holistic view of soil fertility, how crops feed, and how to produce nutritious food. Liebig's reductionism no longer has a leg to stand on. It has been conclusively shown that dependence on chemical fertilizers and pesticides degrades the soil, harms crops, and diminishes food quality. The implications for a new, sustainable world agriculture are enormous.

four

Crop–Nutrient Availability and Deficiency

The primary crop-production strategy for organic growers is to systematically feed the soil animals a quality balanced diet of organic matter and nutritive amendments while providing a habitat for them to mineralize and deliver the nutrients to the plants. We've already discussed the "room" aspects of this room and board equation. Tilth, colloidal aggregation, cation-exchange capacity, carbon/nitrogen ratio, and chelation all have to do with the soil's ability to hold on to, store, and provide nutrients for the plant. Numerous other factors affect how those nutrients become available to begin with, as well as how they can be supplemented when in short supply. Coming full circle, many regular organic practices benefit these processes. Crop rotation, cover cropping, minimum tillage, the use of biostrips, and other soil-fertility enhancement techniques are the overall best management practices for producing sustainable crop yields.

Soil pH

While back in the old days a farmer might grab a handful of soil and taste it to determine whether it was "sweet" or "sour," an electronic pH probe today quickly determines soil acidity and how much lime is needed to release crop nutrients from strong ionic bonds. The test measures the hydrogen ion concentration in the soil solution. Although the scale measures from 1 (extremely acid) to 14 (extremely alkaline) with 7 as neutral (neither acid nor alkaline), soils generally range between the values of 4 and 8. Acid soils are typical in humid regions, while soils in arid climates tend to be alkaline. It is a logarithmic scale: a soil with a pH of 5 is ten times more acidic than one with a pH of 6.

Most vegetable and legume crops growing in a temperate climate prefer

a pH around 6.5, although acid-tolerant plants such as blueberries, potatoes, rye, oats, and alsike clover are common exceptions. Going to 7 and
8 on the alkaline side of the scale tends to make essential elements such as
iron, manganese, and zinc less available to crops. In an acid soil below 5.0
that same manganese and iron as well as aluminum become soluble and
can have a toxic effect on plants.

There are also many synergistic microbe/plant relationships in the root
zone that depend on proper pH. Most of the decomposer bacteria thrive
at a pH of around 6.5. Lower pH values can inhibit their activity, slow-
ing nitrification from organic matter or fertilizers, for example. However,
acid soil conditions help to inhibit the potato scab fungus as well, and
promote the specific acid-loving bacteria responsible for delivering nutrients to blueberry plants.

Spreading calcium in the form of pulverized limestone is the usual
means for raising pH and unlocking the nutrients bound up in acid soils. It
is important to apply lime by analysis; over-liming can produce negative
effects. Too much speeds the decomposition of soil organic matter residue
and humus. Excessive biological activity rapidly consumes and oxidizes
the organic matter, even biting into the bone of the ancient, structural
humus. Proper lime applications release nutrients, stimulating bacteria
populations and plant growth, which in turn contribute more organic
material back to the soil.

Conventional farmers encounter further difficulties with soil pH.
Phosphorus fertilizers, for example, combine acids with rock phosphate to
form superphosphate. The resulting soluble phosphorus may not only run
off as a pollutant into streams and lakes but also acidify the soil, requiring
more limestone applications and further decomposing humus.

Soil Depletion

Practices that don't pay enough attention to the synergy of soil fertility can end up depleting it instead. Imbalances may manifest themselves
as erosion, insect infestations, disease outbreaks, and weed pressure.
Agribusiness treats these symptoms as separate problems to be solved with
high-cost inputs, including pesticides, biocidal fertilizers, and biotechnol-

ogy. Over time, the quality and intrinsic fertility of such soils continue to diminish and, indeed, most are unable to produce crops without increasing amounts of chemicals.

Crops differ in their nutrient requirements for growth, and some plant families are more demanding than others. Among vegetables, corn is a heavy feeder as are brassicas (cabbage, cauliflower, and Brussels sprouts), cucurbits (cucumbers, summer and winter squash), leeks, celery, tomatoes, spinach, and Swiss chard. Some plants are fairly neutral in their demands and actually prefer not too rich a soil. Many herbs fit into this category as do carrots, beets, onions, radishes, and turnips. A third category consists of plants whose growth contributes to the soil. Peas, beans, and other legumes have the ability to fix nitrogen from the air in conjunction with their rhizobium counterparts.

Nutrients also leach away into the subsoil, groundwater, or streams—chiefly through too much rainfall or over-irrigation. Volatilization takes them in a gaseous form into the air. Erosion removes soil particles as well as nutrients. In temperate climates with regular rainfall, however, many nutrients have simply migrated down into the lower subsoil horizons and can be captured by deeper-diving roots and pumped back up into the plant. When the organic residues are tilled into the soil, the nutrients become available for subsequent plantings. Because crop varieties generally have been selected and bred to be shallow-rooted (and coddled and fed by the farmer), they are unable to tap the resources of the lower soil horizons. Cover crops and sod crops, therefore, are important reclaimers and recyclers of migrated nutrients otherwise lost to the agricultural system.

Crop rotations that take advantage of feeding requirements and rooting depths can help rest and replenish the soil. The larger the plant-family diversity in the rotation, the less any particular nutrients will be targeted. Crop trash turned back into the soil will limit the depletion to what has been removed by harvest. The rotations and diversity also help to break insect, weed, and disease cycles. For vegetable growers, a good rule of thumb is to rotate out of the same crop family for three years running. Potatoes shouldn't be followed by another solanaceous crop such as peppers, for example, but by a cucurbit such as winter squash, perhaps, and then in the third year by a root crop such as carrots.

Overly specific rotation schemes, however, can quickly be doomed by bad weather and a host of other vicissitudes that can turn the best-laid planting plans into chaos. Flexibility is a key attribute for successful farmers and gardeners alike. Sometimes you got to do what you got to do to get the crop into the ground, plan be damned.

Deep-rooted cover crops regularly scheduled into the rotation over the winter or for a full season's replenishment are important fallow-time practices. Inexpensive and fast-growing catch crops such as annual ryegrass can be sown to provide cover any time an opening between crops arises, whereupon they catch and hold nutrients in the root zone. Oats are a good fall-planted catch/cover crop for preventing erosion and holding nutrients from leaching away over the winter. Cultivating your own cover-crop consciousness (that is, an elevated personal discomfort level at the sight of bare soil) goes a long way toward preventing the loss of soil and nutrients.

Although the economic pressures of small-scale farming often seem to necessitate cropping land fence line to fence line, building some excess land capacity into a farm plan enables longer-term rotations that move sections out of crops altogether. Sod-based plant communities and perennial legumes left in for a year or more can do more for soil health and productivity than many dollars worth of soil amendments and other inputs. Their growth activity balances the soil, replenishing nutrients as well as regenerating biota and rebuilding soil structure.

Nitrogen

While our planet's atmosphere is around 78 percent nitrogen, this primary nutrient is generally in short supply for crop production. Part of the reason is its basic volatility. Nutrients from decomposing organic matter (from manure plowed in, for example, or a legume cover crop tilled under) are converted to inorganic mineral forms that are available to plants by the soil microorganisms in the process called mineralization. While the nitrogen in living protein is relatively stable, the microbial digestion process converts it to ammonium (NH_4). It is then oxidized in the soil with the help of aerobic bacteria into nitrate (NO_3^-), the form most easily utilized by plants.

Nitrate is highly soluble, however. The problem is ionic: like charges repel. The negatively charged nitrate molecules are repelled by the similarly negatively charged cations in fertile soil—the residual organic matter, humus, and clay particles. While a high cation–exchange capacity is valuable for holding on to positively charged soil nutrients—calcium (Ca^{++}), potassium (K^+), and magnesium (Mg^{++})—and keeps them from leaching with every rainfall (or irrigation), nitrate is actually speeded on its way.

Heaviest feeders (*bottom row*): corn, cabbage, celery, cucumber. Medium feeders prefer less rich soil (*middle row*): carrots, beets, onions. Soil-enriching, light feeders (*top row*): peas, beans.

Losses can also occur through volatilization. If the pH in a compost pile, for example, goes above neutral, the ammonium converts to ammonia gas (NH_3) and escapes into the air. If the compost lacks oxygen, then denitrification via anaerobic bacteria occurs, converting the nitrogen to the pollutants nitric oxide (NO) and nitrous oxide (N_2O). Nitrogen (N_2) is also lost back to the atmosphere.

Fertile soils rich in organic matter with good structural tilth have numerous buffering capacities that bind nutrients and provide them gradually (through the already-discussed chelation capacity and colloidal content, for example) and provide the slow, biologically induced releases of mineralized nutrients that crop plants can utilize over the growing season.

Hunger Signs in Crops

While the essence of organic agriculture is to feed the soil, there's no way around the fact that a portion of your soil's fertility is removed along with each harvested crop. Over time, continuously growing the same crop (including mixed vegetative species such as hay) will deplete some key nutrients. A wet year can leach them away. Increasing soil acidity as a result of polluted rainfall may lock them up. Liebig's law of limiting factors forever looms on the soil horizon.

While there's no substitute for periodic soil tests (every three to five years, or every two to five years on irrigated soils) it pays to develop a practiced eye for seeing what the crops are telling you directly. Although the nutrient requirements of various vegetable crops vary considerably, most plants exhibit readily identifiable symptoms of plant malnutrition—similar signs that point to a particular soil-nutrient deficiency.

Nitrogen Deficiency

Absolute soil nitrogen levels are notoriously hard to test for. Send parts of the same soil sample off to a dozen different soil-testing services and you're liable to get twelve different recommendations, most erring on the side of excess applications of "insurance fertilizer" to guarantee bringing in the crop. Also, samples you take from your cold April soils may yield considerably different values than the same soil tested in July—depending on the weather, organic matter content, and conditions brought about by tillage, compaction, and other farming practices. Testing at the same time of year (midsummer or early fall is best) over a period of several years is a better means of verifying your nitrogen replenishment practices, helping to establish a relative portrait of nitrogen usage. Still, even normally ample nitrogen reserves can be readily leached out of the root zone during wet periods, for example, or become oxidized away by excessive tillage.

One good exception to this rule of thumb is the pre-sidedress nitrate test (PSNT), or Magdoff test, developed at the University of Vermont. Originally developed to determine nitrogen levels for both field corn and sweet corn, it is now being expanded to fit some brassica and cucurbit crops. There are guidelines for potatoes and other solanaceous crops as well. Giving credits for previously incorporated nitrogen-rich manure or legumes, the PSNT samples are taken when the corn is 6 to 12 inches tall to determine more exactly how much more nitrogen should be side-dressed (if any) to provide adequate nutrient levels for the crop.

In the field, lack of nitrogen shows up rather quickly as retarded plant growth, poor color and plant appearance, and low yields. Even with plants that appear healthy, however, underlying deficiencies can appear suddenly later on. Using tomatoes as an example, the soil nitrogen reserves may be adequate to support initial seedling growth, producing a healthy deep green color in the stems and leaves along with vigorous growth. The maturing plants may look fine too, but if the supply of available nitrogen becomes diminished at the fruiting stage, plant growth may dramatically slow down, and it may be too late to do anything to bring in much of a crop.

The first key indicators are the leaves at the top of the tomato plant. They remain small and thin with their tips turning lighter and lighter shades of green to pale yellow. Then as the leaves of the whole plant turn yellowish green, the leaf veins and stems begin to turn purple. The reduced leaf capacity limits chlorophyll production, causing the flower buds to yellow and shed. The stems become hard and fibrous, and beneath the soil the roots become stunted, turn brown, and die off. By now the capacity of the plant to manufacture carbohydrates is severely diminished, dramatically reducing fruit production. The tomatoes that do make it to maturity start off pale, yellow, and small before ripening to unnatural, deep red colors.

Cucumbers exhibit many of the same nitrogen deficiency symptoms— yellowing foliage, stunted growth, and hard stems. The fruits turn a pale, lighter green color and become pointed at the blossom end. When such symptoms are extreme, yields drop off precipitously. However, even a moderate nitrogen deficiency will produce lower harvests.

Organic growers should take into account the nitrogen values of their regular practices—composts, legume cover crops, barnyard manures, plant residues, and so on—that contribute to a residual supply of nitrogen in the form of soil microbes. Seed meals, fish meal, and blood meal, or commercial formulations from North Country Organics or Crop Production Services, are readily available sources useful as a side-dressing for rectifying temporary deficiencies.

Phosphorus Shortage

Phosphorus deficiency usually shows up after land has been in production for a few years. Its symptoms are often more subtle, manifesting as slower plant growth and delayed maturity, as phosphorus also helps plants to absorb other key nutrients. Phosphorus is a component of every living cell and concentrates especially in seeds and the growing tips of plants.

In tomatoes, lack of phosphorus appears as the development of a reddish purple color on the underside of the leaves that often shows up on young plants and continues to maturity. The deficiency develops into pronounced spots and later spreads over the whole leaf, giving it a purplish hue. The veins finally become reddish purple as well.

In the soil, the phosphate ion (PO_4) has a negative charge that does not cling to humus or clay particles, and excess phosphorus (via overapplications of manure, for example) can be lost to surface runoff, which pollutes waterways. Because most forms of phosphorus are insoluble and tightly bound to soil particles, it can become a pollutant when soil erodes. However, most soils usually retain substantial reserves of phosphorus that are tightly held with other elements, unavailable to plants unless mineral-

ized by a healthy foodweb with ample mycorrhizae. Buckwheat is a valuable phosphorus accumulator. Quick-growing summer cover crops gather
residual soil phosphorus and sequester it in the soil's biomass in a form
available to following crops.

Organic growers have several primary sources to choose from. Rock
phosphate is a mined material containing 30 percent phosphate. Only
3 percent is directly available; the rest must be mineralized over time
by the soil foodweb. Bone meal, at 12–18 percent phosphate, is much
more immediately available but is much more expensive. Raw bone meal,
ground up with meat and cartilage residues, also has 5–7 percent nitrogen
content. The more refined steamed bone meal is lower in nitrogen but
easier to store and handle.

The use of slaughterhouse waste products needs to come under new
scrutiny in organic agriculture. Despite outbreaks of bovine spongiform
encephalopathy, or BSE (commonly called mad cow disease), in Great
Britain and Europe, the presence of this disease has not been documented
in the United States.

Potassium Starvation

Potassium (K) is more easily removed from the soil system in the crops
at harvesttime than other nutrients. It is also highly leachable. While an
adequate supply promotes overall plant health and disease resistance, deficiencies show up fairly quickly as reduced plant vigor and retarded growth.
If the deficiency is severe enough, symptoms will show up in the seedling
stage of crops; otherwise the lack may not be noticed until the plants near
maturity, when it is too late to rectify the situation for the crop.

The major hallmark of potassium deficiency is slow growth and stunted
plants. In tomatoes the young leaves become finely crinkled while older
leaves turn a gray-green color and then develop yellowish margins. The
injury progresses toward the center of the leaf, creating large, light-
colored spots that often turn bright orange. A pronounced bronzing of
the leaf tissue occurs until the leaves finally turn brown and die.

This characteristic bronzing effect is very evident in brassica crops,
culminating in brown spots. Carrots exhibit curled leaves first; they then

progress to gray-green and bronze. Cucumbers develop an enlarged tip end on the fruit along with the bronzing, then dying, of the leaf margin.

The most common soil amendment used to rectify potassium imbalances on organic farms is Sul-Po-Mag, a natural rock powder that is 21 percent potassium and 11 percent magnesium. Once the base levels are achieved, soil potassium can be maintained with adequate amounts of organic matter from compost, cover crops, and plant residues. Most manures (except poultry) have ample potassium reserves as well. Slower-releasing potassium sources include greensand and granite dust, generally applied at 2 to 3 tons per acre depending on soil analysis.

While wood ashes are high in potassium and may be easily available, care should be taken in their use. A liming effect, equivalent to about two-thirds that of limestone, could raise the soil pH too high if levels are already adequate. Highly leachable, the ashes should be stored dry over the winter and applied in the spring.

Calcium

Calcium is needed to produce cell walls. Once in place, it becomes immobilized and can't be translocated to a different part of the plant to form new cells. A fresh supply is therefore needed by growing plants to allow the growth of new roots and leaves. It also affects the growth of existing leaves, which will exhibit curling when calcium is in short supply. Plant families require differing amounts—legumes need much more calcium than plants in the grass family, such as corn.

Plants lacking calcium suffer retarded vegetative growth and exhibit thick, woody stems. Tomatoes have yellow leaves on the upper part of the plant while the lower leaves remain green—just the opposite of nitrogen, phosphorus, or potassium deficiencies, where the lower leaves are discolored and the upper leaves and stems remain normal. A calcium deficiency may also be caused by droughty conditions, inasmuch as calcium is transported by water. Blossom end rot on tomatoes indicates a failure of calcium translocation at the fruiting stage, but the remedy may be to increase water reserves through irrigation or mulching rather than adding more calcium.

Deficiency in peas shows up as red patches on the leaves that eventually spread out across the whole leaf. The overall plant gradually turns pale green and then white, causing slow growth and dwarfed plants.

Limestone is calcium carbonate. Carbonate is the constituent that neutralizes soil pH. Over-liming has a detrimental effect by overstimulating soil organisms to rapidly consume organic matter, causing microbe imbalances and soil nutrient losses. If your soil's pH is at optimal levels but calcium levels are low, gypsum (calcium sulfate) can be used without changing pH. It is naturally mined and contains 23 percent calcium and 17 percent sulfur. Gypsum is also valuable for alleviating compaction and promoting drainage in some clay soils—inducing chemical reactions that flocculate (group together) the clay platelets into a granulated structure.

five

The Farmers' Footsteps

Organic agriculture traces its roots through a lineage of accumulated farmer knowledge that dates back to ancient times—knowledge gained from working the soil. What is considered conventional agriculture today in fact represents a distinct departure from the historical mainstream. Industrial methods may have their place in manufacturing "plants" or on the factory floor, but they soon run up against their natural limitations when applied to ecological systems and agriculture. Essentially, organic farming is a biologically positive practice with many beneficent environmental, social, and health effects—one that elevates the soil to its rightful place at the center of agriculture.

One legacy of the Industrial Revolution is the propensity to save money by externalizing the costs of production to the environment, where they show up as pollution, degradation of nature, and negative health effects. While the large mono-cropped commodities (such as corn, cotton, wheat, and soybeans) employ mechanistic efficiencies of scale for tilling, planting, and harvesting, biologically such production is achieved at quite a cost. Beyond the immediate toxicity of chemical fertilizers and pesticides, there are more intractable effects, including the continual losses of topsoil to erosion, the standardization of the seed supply, the need for access to capital, small-scale farmers being forced out of business, and an incredible loss of biodiversity both above- and belowground.

Production agriculture still leaves organic growers relegated to the smaller niches of the food system. Despite the steady growth of the overall organic market—more than 20 percent a year for the past decade—it still represents only a slim slice of the pie. Although some consumers will pay a premium for fresh, local, and organic food, the low basic food prices are still established by distant, large-scale chemical producers. The pay scale for a farmer's own field labor is largely set by the subsistence wages doled out to exploited migrant farmworkers by the agribusiness producers.

Genetically Modified Soil Organisms

Organic growers now have to contend with hazards that are greater than ever, ones that disregard the interactive wholeness of ecosystems in a whole new way. Transgenic technologies are creating novel life-forms across species lines that have never existed before. Briefly tested under laboratory conditions, they are now grown on millions of acres and mass-marketed around the world. The environmental effects are largely unknown. Barely tested at all are the effects biotech crops have on soil life.

Genetic pollution has enormous ecological stakes, as self-replicating organisms are being released willy-nilly into the environment. They can spread through pollen and pollinators to other crops (Bt field corn to organic sweet corn, for example), or to nearby wild, weedy relatives (as with transgenic canola to wild brassicas), where they are free to take on a life of their own. Certified organic crops by definition are subject to economic injury due to pollution from genetically modified organisms (GMOs), with even a farm's organic certification status at stake. In one study, corn pollen was traced from 93 miles away.

The full-time presence of GMOs let loose in the environment poses further problems. *Bacillus thuringiensis*, for example, are natural soil bacteria that were developed as biological pest controls because of their ability to target specific pests in their immature growth stages. The bacterial Bt formulations sprayed on crops by organic growers only live for a few days before being broken down by the sun and weathering elements, producing a minimal impact on other ecosystem functions.

Transgenic Bt crops, on the other hand, are implanted with the concentrated genetic Bt toxin itself (not the bacteria), which becomes expressed full-time throughout all parts of the plant—leaves, stems, pollen, roots, root exudates, seed, and the crop itself. While subsequent testing of transgenic crops has found poisonous effects on non-target species, such as monarch butterflies and green lacewings, little attention has been paid to the impact on soil organisms. The genetically altered root exudates are in a novel, highly concentrated form and are very much at home in their natural soil habitat. Pesticides, herbicides, and chemical fertilizers have already been proven to disrupt and severely degrade soil organisms.

Colonization of soil ecosystems by virulent, self-perpetuating genetically modified organisms could be disastrous.

Add to this the loss of Bt itself as a low-impact pest control, due to increasing pest resistance brought about by its full-time persistence in the field. This will necessitate using more highly toxic chemicals to do the same job.

There's big trouble when agriculture loses touch with the soil. History is rife with civilizations going down or having to move on to greener pastures because of it. The difference today, however, is that we no longer have anywhere else to go.

Search and Research

Organic farmers—and consumers—also receive short shrift when it comes to sharing in the publicly funded agricultural research budget. In a 1997 study, *Searching for the "O" Word*, conducted by the Organic Farming Research Association, less than 0.1 of 1 percent of USDA's research portfolio, both numerically and fiscally, could be considered organic.[1] This not only sets back the advance of beneficial agricultural practices, but also limits career options for scientists interested in doing such research or for students wanting to get into the field to begin with. Publicly funded initiatives that do support organic research include the USDA's Sustainable Agriculture Research and Education (SARE) program and university Integrated Pest Management (IPM) projects. Their budgets remain relatively minuscule, however.

To be sure, systems research presents many challenges to scientists, and it is difficult to assess the qualities of organic agriculture using conventional methodology. It's not surprising, for instance, that organic often comes out dead last in comparison studies using standard side-by-side replicated plantings to test the profitability of conventional and organic crop management systems. In these experiments, the costs externalized to the environment do not enter into the accounting. Nor are the requisite buildup and balance of healthy soil conditions first established. The pesticide treatments employed on the conventional plots serve to drive the pests over to feed on the nearby organic plots. Researchers often fail to grasp that organic is defined not merely by substituting natural inputs for chemical ones, or by what you don't use. Organic certification programs not only mandate a three-year transition out of synthetic inputs, to heal the ecosystem and build the soil, but also require that fields be isolated—protected from spray drift and other adverse effects of nearby conventional practices.

Further barriers to whole-systems understanding are the built-in constraints of the research profession itself. It's difficult for researchers to give up their test plots and dedicate their lab or field allotments to ecosystems approaches. Scientific disciplines have become highly specialized and compartmentalized. Soil scientists and plant pathologists, for example, may have some knowledge of what each other is doing, but they have scant opportunity to work together on joint projects. There is little common ground staked out to test for interrelationships in what they study. Furthermore, the unrelenting pressure to publish research papers in peer-reviewed journals, for career advancement and academic tenure, limits the scope and nature of the projects they take on to begin with. The standard analytical modes are not conducive to studying large numbers of interacting variables.

Despite the difficulties, a new science of agroecology is growing. The understanding of the soil as an ecosystem, a dynamic web of interconnected, interdependent life-forms, requires holistic testing approaches. Recent work by Dr. Larry Phelan at the Ohio Agricultural Research and Development Center provides a good example. To test the relationships between soil management practices, crop nutrition, and susceptibility to pests, he devised a two-pronged approach. First, a three-year study comparing corn crops on neighboring organic and conventional farms found lower levels of serious pests, most significantly European corn borer, on the organic farms. The second phase raised corn under lab conditions in pots in greenhouses using soil taken from both sets of farms. The researchers found that the plants grown in organic soils had less pest pressure and the European corn borers laid far fewer eggs. Although conventional, "high-yield" agriculture often justifies its use of toxic fertilizers and pesticides on the basis of increased production (while disparaging organic systems as low-output) the yields from both the conventional and organic farms were equal.

This project demonstrates the successful integration of the experiential knowledge of farmers with the scientific or empirical approach of the research scientist. These sources of knowledge should be recognized as different but complementary, with each having value. Agricultural researchers should come to recognize the limitations of controlled experimentation conducted on small plots on university land and should be open to learning from the rich observations of farmers. For their part, farmers should play a less passive role in learning how to farm their land by reducing their dependence on fertilizer dealers and land-grant universities.[2]

Agricultural science is developing holistic capacities. Microbiology and genetics—wielded today in the service of corporate hegemony—can cut both ways. Computers connected to high-powered microscopes and DNA analysis are beginning to objectively demonstrate what organic farmers have long known and practiced. As the complex ecological relationships come more clearly into view the reductionist paradigm becomes harder and harder to justify.

Beneath Our Feet

Science has only been able to tell part of the story, however. The animating life forces of the soil—the plants and the farmers, for instance—are left out, unquantifiable. The ancient adage, "the best fertilizer is the footsteps of the farmer," pictures the importance of personal connections to the growing processes and speaks of a heightened consciousness that elevates agriculture into an art. A good grower is said to have a green thumb, a way with plants that transcends methodology. Working directly with the powerful and wondrous elemental life forces resonates in our experience and enhances our actions—and the farm and the garden respond.

Soil-centered holistic practices do have measurable advantages, though. IPM thresholds are based on counts of pests populations: when the pest pressure reaches a certain critical mass, it is time to intervene with pesticides to protect the crop. In organic systems, with legions of beneficial organisms in place above and below ground, conventional intervention thresholds hold no value—a powerful, self-balancing habitat is in place to help maintain crop health.

Most important, nature itself is recast from a hostile force out to destroy and devour a farmer's crops into a powerful and awesome ally. The grower, too, becomes an integral part of the whole system, privy to working with nature in a positive, creative, and meaningful way.

To realize this potential, farmers and gardeners most of all need to develop a holistic consciousness. The life that lies out of sight beneath our feet can so easily remain out of mind. Attention is easily diverted by deer flies buzzing in for a bite, equipment breakdowns, money matters, and 90-degree heat.

Our mind's eye is needed to see the 600 million-plus bacteria teeming in each spoonful of fertile topsoil. How many billions of life-forms lie beneath each footstep as you stride across the field or along the garden? The vision fleshes out and elevates the importance of each and every farm practice: the soil-amendment bulk order, the tons per acre of compost spread, or the acres of cover crops turned in to the soil.

PART 2

Organic Weed Management